Making Maple Syrup

by Elizabeth Moore

Consultant:
Adria F. Klein, Ph.D.
California State University, San Bernardino

capstone classroom
Heinemann Raintree • Red Brick Learning
division of Capstone

Late winter comes.

't is time to make
naple syrup.

The farmer drills a hole in the maple tree.

The farmer puts a tap in the hole.

Sap drips out of the tap.

Sap drips into a bucket.

The sap is collected after a day or two.

The sap is poured into a large cooker.

The sap boils in the cooker.

The farmer stirs the sap.

The maple syrup is almost ready.

Now, the maple syrup is ready.

The syrup is put in bottles.

The bottles go to the store.

Yum! Maple syrup is good!